Jersey's
Geological Heritage

- Sites of Special Interest -

Ralph Nichols &
Samantha Blampied

- 2016 -

States
of Jersey

Jersey's Geological Heritage: Sites of Special Interest

First published in Great Britain
in 2016 by Société Jersiaise
www.societe-jersiaise.org

A catalogue record of this report is available from the British Library

ISBN 978-0-901897-48-0

Contents

Foreword

The rocks that outcrop in Jersey are among the oldest on the planet and are relatively well-exposed, especially in coastal locations. The age and nature of the rocks makes them of particular interest to scientists and some locations are recognised as being of international importance.

In 1989, the British Geological Survey (BGS) published a detailed description of the geology of the island of Jersey to accompany a geological map published in 1982. Jersey was considered to be one of the BGS's Classical Areas of British Geology and the publication summarised and built upon extensive previous work carried out by numerous workers. It also drew attention to specific localities where notably important geological features could be observed.

During the 1990s, considerable effort was put into the cataloguing and documentation of geological sites of particular local, national and international importance, with the specific characteristics of each site being mapped and described in detail. This phase of research was presented in an unpublished report in 1996, recommendations from which led to the 22 sites of greatest importance being designated as Sites of Special Interest (SSI) and protected under the Building and Planning Law, 2002.

The present publication therefore represents the culmination of extensive work carried out by numerous researchers over a considerable period, to provide a comprehensive account of Jersey's 22 geological and geomorphological SSIs. This publication provides clear information, including detailed geological descriptions and interpretation, together with photographs for each of the SSIs.

The authors are to be commended for producing a readable and informative guide to the most important geological sites on the island of Jersey. This guide will undoubtedly advance the future enjoyment and understanding of geological and erosional aspects of the island. This publication will also provide a more widespread recognition of the importance of these unique sites and promote the preservation and protection of them, not only for the future enjoyment of interested Islanders and visitors but also for future study by geology students and specialists.

Colin S Cheney CGeol, MSc, FGS, BSc
Chairman, Geology Section (Société Jersiaise)

Preface

The establishment of a network of Geological Sites of Special Interest (SSIs) on Jersey was initiated by Gerard Le Claire, Chief Officer of the Environmental Services Unit (ESU) of Planning and Building Services, Jersey. This was undertaken in order to document and protect those sites that are the most important examples of Jersey's geological history. The sites were chosen by Dr John Renouf, a Jersey research geologist, who worked with Debbie Davis on this project over a prolonged period. In 1996, Debbie produced a report for the Department of the Environment which contained detailed research and recommendations. From this work, twenty-one sites were selected for Geological SSI status and their justification given in detail. This guide covers all these sites plus the St Ouen's Bay peat beds.

The Jersey geological sites were assessed according to international guidelines for Sites of Special Scientific Interest (SSSIs) and classified into sites of International or Local (A, B or C) significance. This was based on their recognition by academic and consultant research geologists from Jersey, the United Kingdom and France, many of whom had published results from their Jersey studies in national and international journals. The sites were named then described and illustrated on maps by identifying core features for protection with a surrounding buffer zone. The reasoning for their protection was also given, and all the information provided justification for their designation as appropriate sites.

Gerard Le Claire gave a copy of the final report to Dr Ralph Nichols of the Geology Section of the Société Jersiaise for comment and amendment. This was done and contact was made with Mike Freeman, Former Principal Ecologist and acting Chief Officer of the ESU, who had taken over following Gerard Le Claire's tragic death in a helicopter accident in 2001.

The present authors gratefully acknowledge Mr Freeman's suggestion that we work with Lindsey Napton who was responsible for processing the information into an acceptable form for the States of Jersey to pass the necessary law. The sites were visited and the SSI boundaries established using a GPS. Lindsey Napton then contacted the landowners to obtain permission for their land to be designated an SSI. Dr Nichols then compiled a brief report for each site using the same headings and listing the key features but added photographs to illustrate their geological significance and to record their present condition. During this process two other SSI sites were added to further illustrate Jersey's geological heritage.

With a view to the publication of a guidebook of Jersey Geological SSIs for the public, gratitude is expressed to John Pinel, the current Principal

Ecologist and Assistant Director (Natural Environment), who agreed to a final period of editing and photography which was co-ordinated by the Department's Research Ecologist, Nina Cornish, with the expert help of David Tipping during Lindsey Napton's leave of absence. The technical geological descriptions for each site and photographs of key features were amended and presented for easier understanding by Samantha Blampied, an undergraduate BSc student on a placement project with the Department. Finally, a debt of gratitude is owed to Dr Paul Chambers for his patience, guidance and expertise in all the publication phases of this work, and to Roger Long of the Société Jersiaise for the final editing.

Ralph Nichols, 2016

Introduction

The geology of Jersey is significantly different to that of the United Kingdom and even from that of the other Channel Islands. The island has always been of interest to British geologists but many of its individual geological sites are of regional and international significance and have attracted attention from across the world. In recognition of this the States of Jersey has designated 22 of the island's most important geological outcrops as Sites of Special Interest (SSIs) so that they may be protected from development and preserved for future public enjoyment and research purposes.

This booklet offers an introduction to all of Jersey's geological SSIs with the aim of promoting knowledge of their existence and importance to Jersey's residents and tourists. All the geological SSIs are listed here along with information about their location and a basic description and photographs which highlight the sites' significance and their salient features. More detailed scientific information for most of the sites may be found in the references listed on page 11 but especially useful are: Bishop and Bisson (1989); Keen (1993); and Nichols and Hill (2004).

The geological SSIs are protected under the Building and Planning Law (2002) and should not be interfered with or disturbed in any way. Hammering, digging and the collection of samples from outcrops is prohibited and care should be taken to stick to existing footpaths to avoid excessive erosion. Thumbnail aerial photographs in this booklet indicate the approximate extent of the SSI boundaries but these are for guidance only. When on the ground it may be difficult to discern the limits of an SSI as its geological features may blend into the surrounding landscape.

Further information on SSI boundaries, the SSI scheduling process and its implications may be obtained from the States of Jersey planning website (www.mygov.je/Planning/Pages/Planning.aspx) where full documentation for each SSI (including maps) may be found in the 'Natural Sites' section.

It should be noted that many of Jersey's geological SSIs are situated in locations that have uneven ground, unstable surfaces, steep drops and unpredictable tides. In recent years there have been many deaths, serious injuries and dramatic rescues by unwary locals and tourists who have set out to explore the island's more remote coastal and inland locations.

All of Jersey's geological SSIs (but especially the coastal cave sites) should be viewed as hazardous with an associated risk of injury, and extreme caution is advised at all times.

It is particularly important that people do not visit the sites on their own and when there, keep away from cliff edges, overhanging rock and unofficial footpaths. Sites on or near the intertidal zone should only be visited on a falling tide and it is advised that people wear appropriate clothing and footwear and carry a mobile phone, water and first aid kit. Advice concerning individual sites may be obtained from the Department of the Environment.

The 22 sites scheduled as geological SSIs represent all aspects (and ages) of Jersey's geology but this list is by no means definitive and the island has many other sites that are of scientific and aesthetic importance. Please act carefully and considerately around all areas of Jersey's natural environment so that it may be conserved for its flora and fauna and the future enjoyment of others.

A Brief Geological History of Jersey

Jersey's rocks generally fall into one of two categories. There are the rocks which are all over 400 million years (Ma) old and of a sedimentary and igneous nature; and there are the semi-consolidated rocks which are less than 2 Ma and are all sedimentary. Below is a brief chronological history of Jersey's geological evolution which may help to put the significance of some of the SSIs in context; a basic geological map of the island is shown on pages 4-5. Further information may be obtained from the resources listed on page 11.

Jersey's oldest rocks are the grey, bedded siltstones, shales and sandstones (greywackes) of the Jersey Shale Formation (660-590 Ma) which crop out from St Ouen's Bay through the centre of the island to St Aubin's Bay, and north of St Helier to Gorey Harbour beach. They represent ancient sediments deposited by submarine currents in an offshore deltaic environment.

The shales are overlain (disconformably) by two different types of volcanic rock both formed around 533 Ma. The first is the St. Saviour's Andesite Formation, which crops out from St Helier north to the Trinity coast at La Belle Hougue headland. This is overlain by the St John's Rhyolite and Bouley Rhyolite Formations which occur from the north to the east coasts of the island.

Around 590 Ma two large land masses (plates) approached from the north-west and south-east and then collided with each other, uplifting all the above rocks, folding and faulting them in the process. A series of dioritic and granite masses were intruded into the overlying rocks creating the South-east Granite complexe (c. 570 to 509 Ma) then the South-west and North-west Granite complexes (c. 550 to 483 Ma; and 465 to 426 Ma respectively).

Uplift, weathering and erosion followed, creating mountains within which were pebbly flash flood deposits that became the Rozel Conglomerate Formation (427 Ma) of north-east Jersey. With the mountain building came a period of minor intrusions which saw the emplacement of pink-red aplite veins and larger, dark grey dolerite and brown mica lamprophyre dykes. These form the Jersey Main Dyke Swarm (440 to 425 Ma) which is best exposed on the seashore east of St Helier.

Starting about 400 Ma the island became tectonically quiet and instead of accumulating rocks, it was subject to millions of years of erosion and weathering. It is this that produced the island's characteristic flat-topped profile and which, during times of low sea levels, allowed rivers to carve wide valleys off the Normandy and Brittany coasts.

When sea levels were higher these valleys flooded and isolated the series of flat-topped plateaux that today form the individual Channel Islands. Seabed outcrops of Eocene (55 Ma) and Pliocene (1.8 Ma) sediments and limestones suggest that the current topography of the Channel Islands was probably in place at the end of the Mesozoic Era (65 Ma) and possibly much earlier than this.

During the Pleistocene (1.8 Ma to 11,000 years) glaciers advanced and retreated across Europe causing the sea level to rise and fall dramatically. At the height of the ice ages Jersey was part of a barren and frozen landscape that was many miles from the coast. Thick head deposits formed across the island on top of which were deposited thick layers of loess, a type of yellow, wind-blown silt.

During times of glacial retreat, the sea level would rise and surround Jersey creating three fossil beaches, which are currently raised above the high water mark along the coast. Periodically ice age fauna and flora would be attracted to the island including mammoths, woolly rhinos, Palaeolithic Neanderthal and early modern humans. At the end of last ice age (13,000 years ago) the sea level rose until, around 8,000 years ago, Jersey was an island once more. Mesolithic hunter gatherer humans were already resident by this time and around 6,000 years ago Neolithic farming began. In the millennia that followed the landscape has been progressively modified by humans up until the present day, resulting in the modern island of Jersey.

Further Reading

Bishop, A. C. and Bisson, G., 1989. *Classical areas of British geology: Jersey,* HMSO

D'Lemos, R. S., Strachan, R. A. and Topley, C. G., 1990. *The Cadomian Orogeny.* Geological Society Special Publication No. 51.

Institute of Geological Sciences, 1982. *Classical areas of British geology: Jersey. Channel Islands, Sheet 2.* IGS. 1: 250,000.

Jones, R., Keen, D., Birnie, J. and Waton, P., 1990. *Past landscapes of Jersey: Environmental changes during the last ten thousand years.* Société Jersiaise.

Keen, D. H., 1993. *Quaternary of Jersey Field Guide.* Quaternary Research Association

Mourant, A. E., 1961. *The minerals of Jersey.* Société Jersiaise.

Nichols, R. A. H. and Hill, A. E., 2004. *Jersey Geology Trail.* Private publication

Websites

Jersey Geology Trail: **www.jerseygeologytrail.net**
Ice Age Island: **iceageisland.wordpress.com**
Société Jersiaise: **www.societe-jersiaise.org**
States of Jersey: **www.mygov.je/Planning/Pages/Planning.aspx**

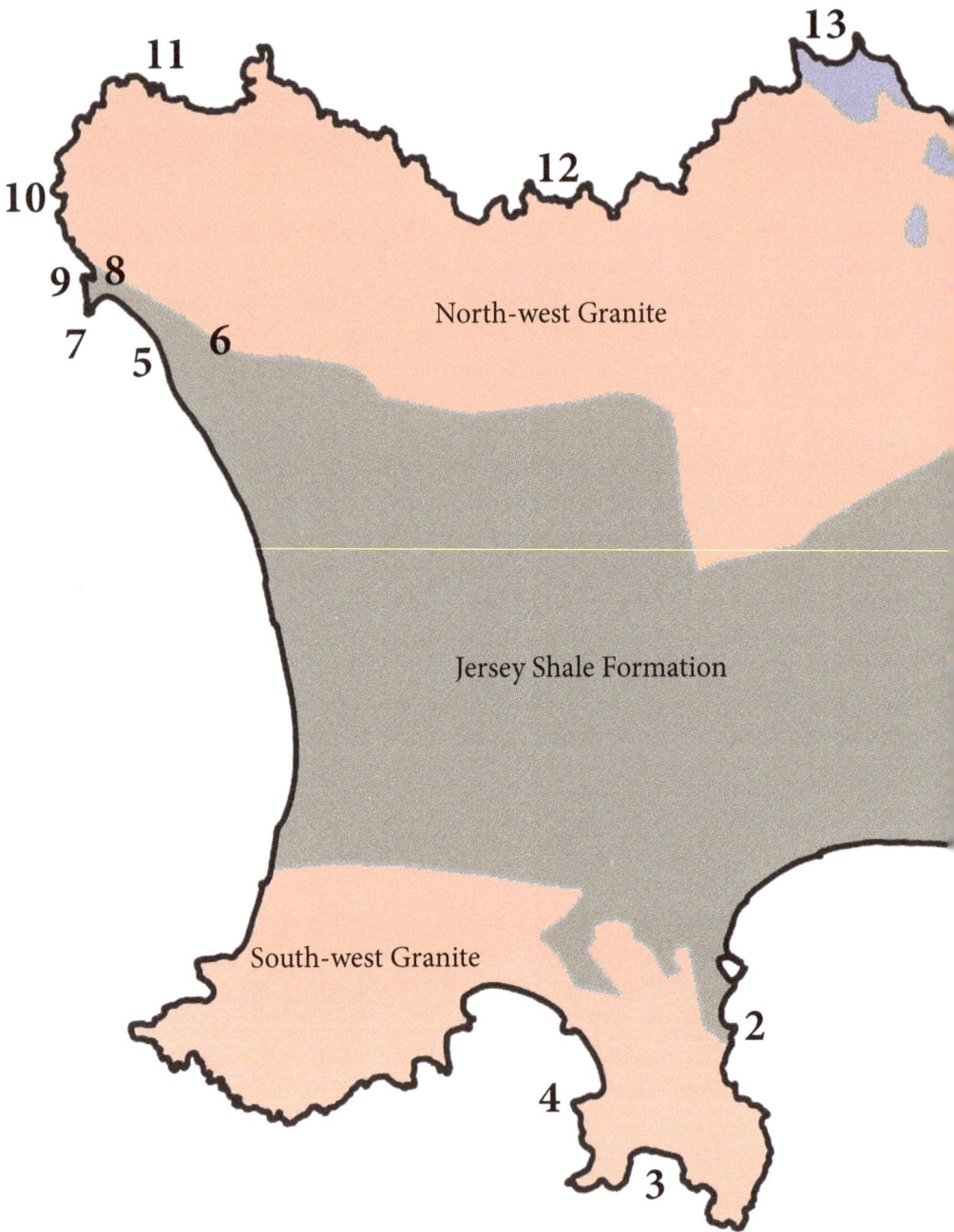

North-west Granite

Jersey Shale Formation

South-west Granite

1 - South Hill; 2 - Belcroute Bay; 3 - Portelet Bay; 4 - La Cotte de St Brélade;

5 - St Ouen's Bay Peat Beds; 6 - Le Mont Huelin Quarry; 7 - Le Petit Étacquerel;

8 - Le Grand Étacquerel; 9 - Le Pulec; 10 - Le Pinacle; 11 - La Cotte à la Chèvre;

12 - L'Île Agois, Crabbé; 13 - Sorel Point; 14 - Giffard Bay;

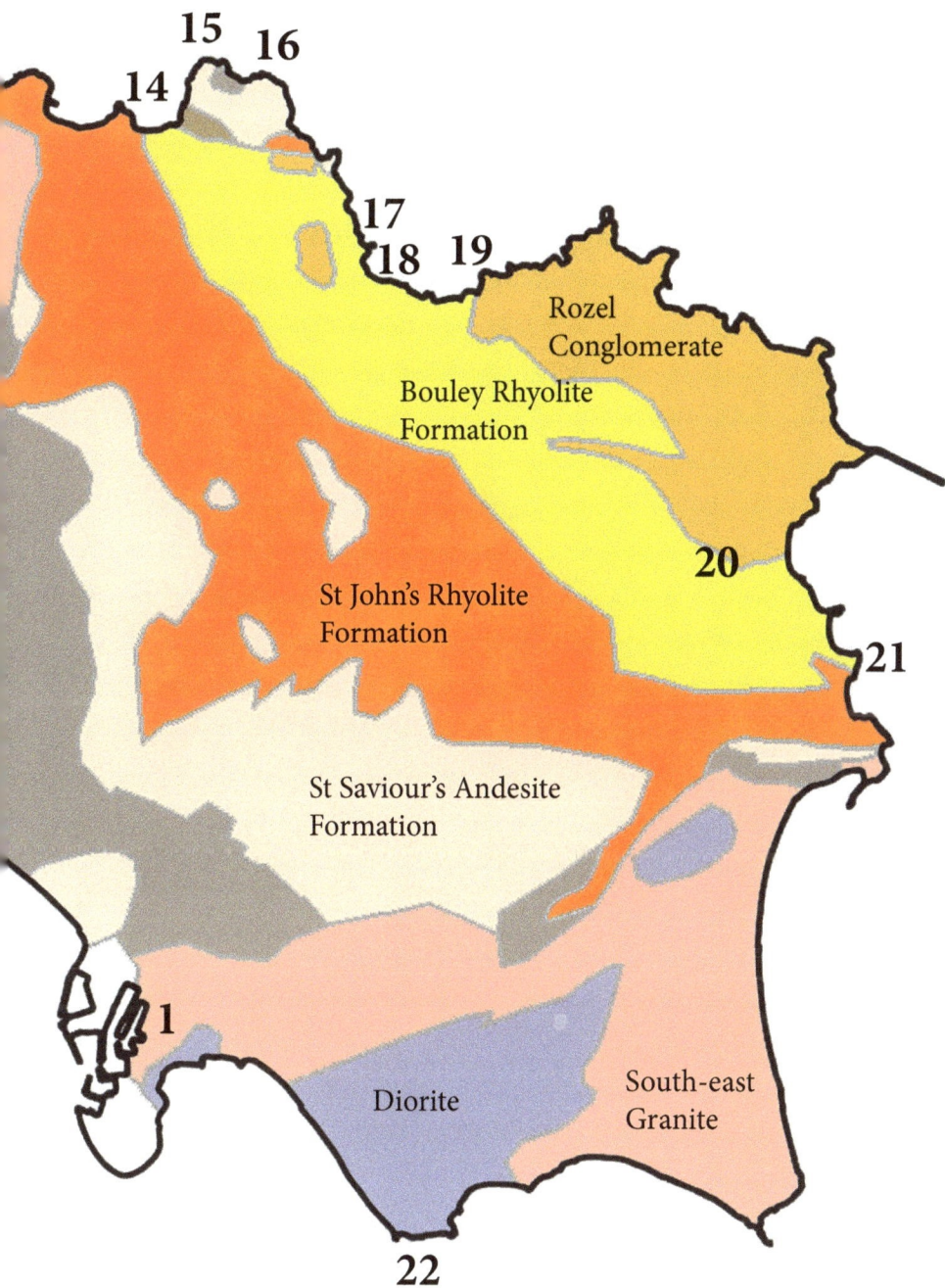

15 - La Belle Hougue Caves; 16 - Les Rouaux; 17 - Les Hurets, Bouley Bay;

18 - L'Islet, Bouley Bay; 19 - La Tête des Hougues;

20 - La Solitude Farm; 21 - Anne Port Bay; 22 - La Motte, Le Nez and Le Croc

1 - South Hill

Raised beach/SE Granite	Pleistocene/Palaeozoic
SSI Status: Local A	Grid Ref: OS 6510 4770

A 40-metre raised beach which consists of a mixed collection of pebbles, mostly of local granite, embedded in sand, can be found under the (now overgrown) gardens at the east of the quarry. A recent discovery (above) crops out on the north-west side of the hill.

In addition, adjacent mica lamprophyre dykes and a dolerite sill can be seen in the Fort Regent granophyre, part of the South-east Granite, and a classic soil creep structure is seen at this site. (Main photo: Tim Liddiard)

A mica lamprophyre dyke in the granophyre which shows an altered margin, is found on the south-east side and strikes north-west to south-east.

A dolerite sill-like intrusion in the granophyre is also visible on the south side of the hill.

The effect of soil creep is well displayed in some of the exposures on the north side, near the car park.

2 - Belcroute Bay

Loess/Dykes	Pleistocene/Palaeozoic
SSI Status: International	Grid Ref: OS 60704810

The succession of Pleistocene soft rock deposits north of Belcroute slipway is one of four main sites on Jersey where interglacial and glacial sediments occur. The periglacial head and loess are thought to belong to the Wolstonian Glacial. The whole succession rests on a platform eroded into the Jersey Shale Formation at the 8-metre level. Dolerite, rhyolite and lamprophyre dykes are found to the north and south. The picture above shows loess and head above the 8-metre raised beach (under land slip).

There is evidence of a raised beach of large granite pebbles and cobbles in a marine sand deposit at the base of the cliffs.

A rare flow-banded rhyolite dyke (one of just three in the island) occurs in low cliffs of Jersey Shale in the adjacent beach to the north.

A mica lamprophyre dyke striking west-north-west in the Jersey Shale Formation in a beach outcrop in the north of the bay.

17

3 - Portelet Bay

Loess/ Raised beach/SW Granite/Sill	Pleistocene/Palaeozoic
SSI Status: Local A	Grid Ref: OS 6070 4710

Portelet Bay is backed by a cliff of partly-consolidated Pleistocene deposits which include head, loessic head and loess overlying a raised beach at the 8-metre level. There are excellent outcrops of dolerite dykes and eroded gullies, a rare sill, and a tombolo spit linking the beach area to L'Île au Guerdain, as well as an outcrop of the coarse Corbière Granite of the South-west Granite with inlets eroded along dolerite dykes (above picture).

An Ipswichian 8-metre raised beach of granite pebbles over lower sand, seen in the cliffs at the back of the bay.

Loess and head above the raised beach. In parts, the loess contains calcareous concretions, a few of which contain fossils of non-marine molluscs.

This dolerite sill, found in the western cliff, is one of only a few sills to be found in Jersey.

4 - La Cotte de St Brélade

Raised beach/Fossil site/SW granite	Pleistocene/Palaeozoic
SSI Status: International	Grid Ref: OS 5933 4750

This famous rock shelter is in an outcrop of porphyritic granite, part of the South-west Granite. It is Jersey's most important Palaeolithic site with fossils and tools very closely linked with its geology. Deposits of an 18-metre and 8-metre raised beach are found here. The rare, continuous range of Quaternary age deposits; basal loess, followed by peat then by glacial head, indicate a change from glacial to interglacial and back to glacial conditions. The oldest deposits are considered to be more than 200,000 years old.

The entrance to the rock shelter at La Cotte de St Brélade. Differing joint patterns show how the rock shelter was formed. (Photo: Jeremy Percival)

Excavations of the loess and head have uncovered a rich harvest of Palaeolithic tools, mammal remains and a few human teeth and bones.

A team from University College London is re-examining the loess and head for artefacts and fossils in their context.

5 - St Ouen's Bay Peat Beds

Peat beds/Fossil site/Artefacts	Holocene
SSI Status: International	Grid Ref: OS 5530 5425

In St Ouen's Bay there are excellent exposures of peat containing tree stumps, root systems, branches and twigs, often in their growth position. The peat is between 5,000 and 7,000 years old and is known locally as the 'submerged forest'. The stumps are mostly from birch and alder trees. The peat is of great archaeological interest as it contains fossil hoof prints, cattle bones, pottery and flint artefacts. It occurs under much of St Ouen's Bay but is usually buried under the sand and only exposed after storms.

A fossilised tree stump and its root systems from north of La Saline. These trees grew in a marshy environment located behind a coastal sand barrier.

Rootlets in sandy beds below peat at Le Port.

Inset picture: A frond of sedge (*Carex* sp.) in the peat.

Twigs in rippled peat beds at Le Port.

Inset picture: A fossil hoof print at Le Port. If found, then features like this should not be interfered with.

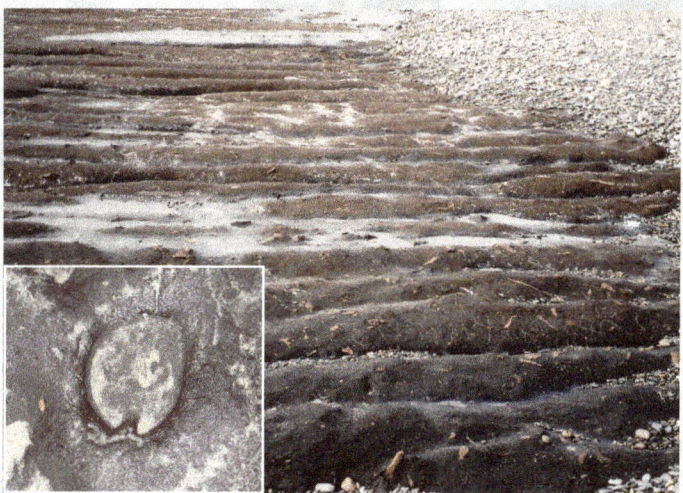

6 - Le Mont Huelin Quarry

Jersey Shale Formation/NW Granite	Precambrian/Palaeozoic
SSI Status: Local C	Grid Ref: OS 5520 5448

At this site there is a sharp and definite contact between the Jersey Shale Formation and the North-west Granite which is easy to locate and examine. Well exposed intrusive contacts such as those involving granites and sedimentary rocks, which are not very common, are present at the quarry. The sediments show clear evidence of bedding and there are curved joints parallel to the junction and master joint planes in the granite. Some molybdenite is also present.

Irregular contact between Jersey Shale Formation beds and the North-west Granite (on the right of the photo). This is indicated by the colour change in the rocks.

A shale inclusion in the granite alongside the contact (behind the branch). The bedded Jersey Shale shows contact metamorphism to hornfels.

A molybdenite and quartz vein in the granite.

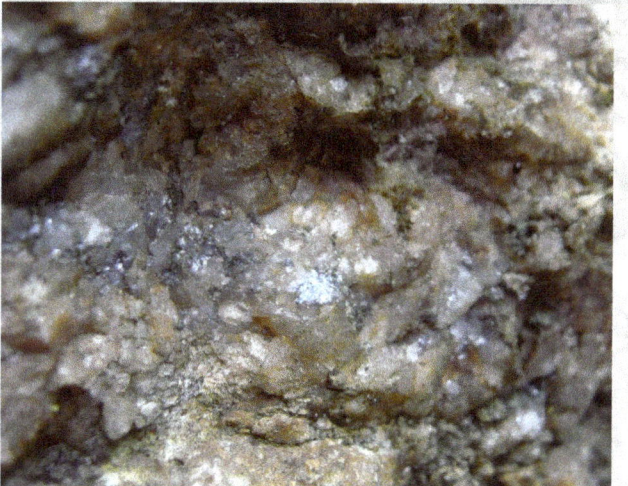

7 - Le Petit Étacquerel

Jersey Shale Formation	Precambrian
SSI Status: Local B	Grid Ref: OS 5481 5466

The north-western area of St Ouen's Bay spans the contact between the North-west Granite and the Jersey Shale Formation. Seen here are excellent examples of two former stacks that were eroded during a period of higher sea level. Various excellent sedimentary structures in outcrops of the Jersey Shale Formation also occur here. Some examples of these structures are shown here and on page 27.

These flow-ripples are sedimentary structures formed on sloping seabeds when sediment-laden turbidity currents flow across the surface.

These thin beds of sediment at an angle to one another are cross-beds. They can tell geologists much about the ancient environment in which they formed.

During deposition, sediment may scour underlying beds or truncate them forming cut and fill structures and erosion surfaces.

8 - Le Grand Étacquerel

Jersey Shale Formation	Precambrian
SSI Status: Local B	Grid Ref: OS 5480 5470

The variation of bedding and sedimentary structures seen here are typical of the Jersey Shale Formation and are beautifully exposed between the slipway at Du Sein and the seawall. The beds exhibit slump-fold structures which were formed immediately after deposition. Sedimentary structures such as flow-ripples and cross-bedding are seen in a variety of sandstones and greywackes to the north of the site.

Sole markings are features that occur on the undersurface of a bedding plane. They are formed when the water currents erode a hollow in the surface of a bed into which new sediment is later deposited.

Cross bedding occurs where thin beds of sediment lie at an angle to one another. The direction the beds are dipping indicates the direction of sediment transport.

Boudinage occurs where sediments are slumped and folded and separated into various lens shapes. Boudinage derives from the French word *boudin*, meaning sausage.

29

9 - Le Pulec

Jersey Shale Formation/NW Granite	Precambrian/Palaeozoic
SSI Status: Local B	Grid Ref: OS 5472 5494

The striking exposure of the contact between the Jersey Shale Formation and the North-west Granite is seen at Le Pulec. There is a sphalerite-galena-ankerite vein associated with the contact and there are excellent exposures of minor intrusive felsite dykes and veins from the granite. Geomorphological features also occur here, including caves and gullies at the 8-metre level filled with mixtures of raised beach pebbles, head and loessic head.

This transgressive quartz and feldspar vein has been intruded along different fracture planes in the rocks.
The inset picture shows iron pyrite crystals in a quartz vein in the shale.

This vein within the shale is formed of sphalerite (zinc blende) of non-commercial value, within the mineral ankerite. The inset picture shows a similar vein in closer detail.

Two examples of a raised beach can be seen alongside the slipway at this site. This photograph shows a raised beach deposit (8 metres above current datum) in one of the gullies at Le Pulec slipway.

10 - Le Pinacle

NW Granite/Sea level features	Palaeozoic/Pleistocene/Holocene
SSI Status: Local A	Grid Ref: OS 5442 5545

Le Pinacle, when viewed from the south, is a striking example of a former stack. It is a remnant of the North-west Granite, linked to the former cliff line by a col (a land bridge) presently being eroded into a cave. There are also examples of Pleistocene 8-metre and 18-metre wave-cut notches from former sea levels of Ipswichian/Eemian and pre-Ipswichian ages at circa 120,000 bp and 200,000 bp respectively. Rare dolerite sills occur in adjacent cliffs.

The top of the former stack, eroded from the NW Granite, is linked to the coastal cliff path by a col which can be reached by a path down the fossil cliff.

(Photo John de Carteret)

Adjacent to it is a rare example of a dolerite sill (or low-angled dyke), one of several in the adjacent cliffs.

On the col there are remains of Neolithic ramparts and tool making. There are also the remnants of an internationally known Gallo-Roman temple.

33

11 - La Cotte à la Chèvre

NW Granite/Sea level features	Palaeozoic/Pleistocene
SSI Status: International	Grid Ref: OS 5525 5668

La Cotte à la Chèvre is an excellent example of a sea cave and a wave-cut notch eroded at the 18-metre raised-beach level in the coarse North-west Granite (438 Ma) with aplite veins. It is also a former Palaeolithic habitation site which has revealed flints used during the last glacial period (Devensian). (Photo Kevin McIlwee)

The cave has a rounded outline and the walls have been smoothed by marine erosion. In the cave there are boulders, pebbles and sand overlain by remnant loess. (Photo: J. Percival)

The coastal cliffs have been eroded by the sea along master joints and faults to produce striking, steep-sided inlets and stacks.

The granite cliffs have been eroded from the coarse outer granite of the NW Granite. They are well-jointed and have been intruded by several microgranite (aplite) veins.

12 - L'Île Agois, Crabbé

NW Granite/Sea level features	Palaeozoic/Pleistocene
SSI Status: Local C	Grid Ref: OS 5960 5570

Various faults, joints and wave-cut notches are clearly exposed at this site. Cliff and inlet landforms have been formed by the marine erosion along faults, joints and dykes. 8-metre raised cliff-cut notches are the main notable features of Quaternary marine erosion. A biotite microgranite, which is a rare variety of the North-west Granite, provides the backdrop for the striking erosional elements of the area. L'Île Agois is also the site of a Dark Ages eremitic settlement.

L'Île Agois has several prominent erosional features including an 8 metre wave-cut notch, master joint erosion and an 8-metre cave.

The cave has been eroded along a narrow vertical dolerite dyke.

A rare biotite microgranite occurs in the in the porphyritic North-west Granite and is exposed in the cliff path.

13 - Sorel Point

NW Granite/Diorite	Palaeozoic
SSI Status: International	Grid Ref: OS 6117 5708

The Sorel Point igneous complex shows a wide variation of different magmas intruded into a small space over a short time period. The exceptionally rare exposures are evidence of how a sub-volcanic magma chamber has evolved. Present are granite, granodiorite, gabbro and diorite rocks formed at successive stages by intrusion and magma mingling. Sorel Point is internationally recognised for its igneous geology. The photograph above shows red microgranite in grey diorite.

Granite dyke and lenses in diorite.
Inset picture: feldspar and hornblende lens in diorite.

Light-coloured diorite column (pipe) in darker-coloured diorite.

Twinned feldspars the porphyritic part of the NW granite.
Inset picture: onion weathering of jointed diorite at the start of path.

14 - Giffard Bay

St John's & Bouley Rhyolite Formations/ Raised beach features	Palaeozoic/Pleistocene
SSI Status: Local A	Grid Ref: OS 6480 5600

This site has the greatest variety of volcanic rocks in Jersey and many are unique to the bay, such as the Frémont Ignimbrite and Giffard Rhyolite. Rare lamprophyre dykes are seen here along with well-exposed fiamme, flow-banded and rheomorphic ignimbrites, and tuffs. It is one of few examples of a calc-alkaline volcanic centre in the Cadomian orogenic belt of the Armorican Massif. An example of the 8-metre raised beach occurs at the back of the bay.

Flattened and streaked ash (fiamme) can be seen at the western end of the bay.

A curved and faulted mica lamprophyre dyke is also found in the west/central part of the bay.

A close up of the 8-metre raised beach showing jasper and rhyolite pebbles embedded in silt.

15 - La Belle Hougue Caves

Raised beach/Fossil site	Pleistocene
SSI Status: International	Grid Ref: OS 6550 5642

An 8-metre raised beach is found in the caves at this site. Belle Hougue Point (above picture) is where the caves are located. Remains of rare fauna are found here, such as bones of red deer, hare and other small mammals, preserved due to calcium leaching from acid igneous rocks. Deposits in the caves have been dated to the Ipswichian interglacial. Mollusc remains present in the stalagmites indicate sea temperatures were around 3°C higher than they are today.

The approach and entrance to the cave at La Belle Hougue are a more dangerous places than other cave sites and should only be visited with an experienced guide.

The entrance and interior floor deposits of the upper cave.

Recently found Pleistocene faunal remains are still present in the finer-grained deposits. (All photos by Jeremy Percival)

43

16 - Les Rouaux

Jersey Shale Fm./St Saviour's Andesite Fm	Precambrian/Palaeozoic
SSI Status: Local A	Grid Ref: OS 6505 5630

Sedimentary structures and deformation features, such as major strike-slip faults in the Jersey Shale Formation are seen at Les Rouaux, as well as an important and rare transition from sediments to a volcanic agglomerate of the St Saviour's Andesite Formation. The North-west Granite intrusion is clearly faulted against sedimentary and volcanic rocks and a dolerite dyke can be seen in the eroded platform. Well exposed erosional features and glacial head are found in the cliffs.

Strike-slip fault trace and eroded north-east to south-west striking dolerite dyke. Inset picture: faulted granite and sedimentary contact.

Agglomerate in the St Saviour Andesite Formation within andesite, with porphyry and shale breccia along the cliff path.

An 8-metre wave-cut platform with wave-cut notch occurs along the back of the bay.

17 - Les Hurets, Bouley Bay

Bouley Rhyolite Formation	Palaeozoic
SSI Status: International	Grid Ref: OS 6688 5463

Seen here are the ignimbrites and rhyolite of the Bouley Rhyolite Formation. To the south of the jetty are excellent green and purple rhyolite ignimbrites. Les Hurets crags are found to the north of the bay (above picture) and seen here are spectacular spherulitic rhyolites (1 mm to 10 cm). Examples are shown opposite and on page 47.

Large oval spherulite with alternating iron rich bands. (Photo: Jo Sonnex)

Uniform spherulites, cavities and irregular banding. Inset picture: uniform silica spherulites in flow-banded rhyolite.

Beds of spherulites of different sizes and shapes and oval, silica spherulites in bedded rhyolite.

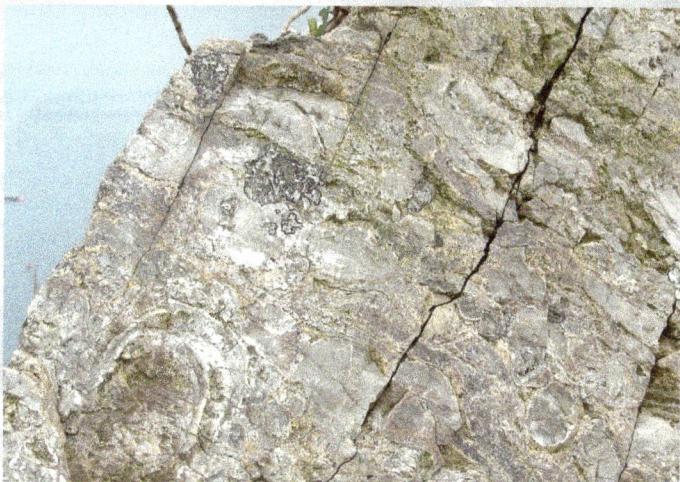

18 - L'Islet, Bouley Bay

Bouley Rhyolite Formation/Loess	Palaeozoic/Pleistocene
SSI Status: Local C	Grid Ref: OS 6716 5440

Excellent fiamme (flattened pumice) texture is seen in the Bouley Rhyolite Formation. Shown in the picture above is L'Islet with loess over ignimbrites. Grey and purple, flow-banded and flattened and streaked ignimbrites crop out below the loess. West of L'Islet are upper green, spherulitic breccias. Also found here are preserved Pleistocene loess and glacial head.

Flattened pumice (fiamme) layers from ash flows.

Flow folds in possible rheomorphic ignimbrites. Inset picture: an isolated flow fold can be seen below the south side of L'Islet.

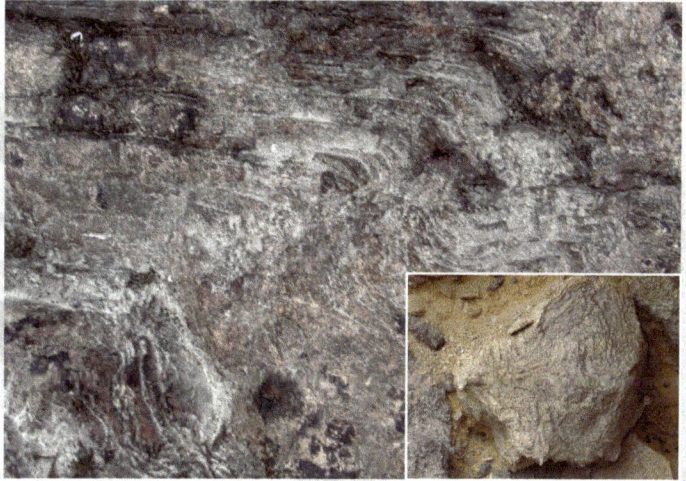

Single and compound spherulites, some with rims. Inset picture: cut specimen of flattened spherulites with silica and iron.

49

19 - La Tête des Hougues

Rozel Conglomerate Formation	Palaeozoic
SSI Status: International	Grid Ref: OS 6792 5444

The best contacts between the volcanic rocks of the Bouley Rhyolite Formation and the flash flood of the Rozel Conglomerate Formation occur in the bay below La Tête des Hougues. There are well-exposed red mudstones with a variety of sedimentary and depositional structures of a playa-sand flat environment and excellent exposures of large and small boulder conglomerates. Basal conglomerate beds are found above the siltstones (above picture). Small and large spherulites and flow folds are seen in the rhyolite.

Large spherulites in flow-folded rhyolite of the Bouley Rhyolite Formation. Inset picture: Isolated quartz spherulite, seen below the cliffs.

Mud cracks and other sedimentary structures in basal red-brown, fine grit siltstones.

Erosional unconformity the unsorted of Rozel Conglomerate overlying Bouley Rhyolite Formation.

20 - La Solitude Farm

Rozel Conglomerate Formation	Palaeozoic
SSI Status: Local C	Grid Ref: OS 7042 5202

At this site are excellent outcrops and cross-sections of basal siltstones, sandstones and cobble beds at the base of the Rozel Conglomerate Formation (above picture). The basal sedimentary structures indicate an early environment of playa-sand flat at the base. La Solitude Farm allows easy access for the study of the geological history of the Rozel Conglomerate Formation from the early sandy basal beds to the overlying flash-flood, unsorted pebble beds.

The Rozel
Conglomerate
Formation with
its unsorted
pebbles and
cobbles.

Pebble beds
in cut and
fill structures
above
siltstones.

Bedded
sandstones
overlying
siltstones.

St John's & Bouley Rhyolite Formations	Palaeozoic
SSI Status: Local A	Grid Ref: OS 7130 5110

From Anne Port Bay northwards, the Anne Port Ignimbrite (an ash unit of the St John's Rhyolite) displays excellent eutaxitic and parataxitic textures (flattened pumice and shards), and is the best example of the upper section of St John's Rhyolite. Excellent rhyolite lava flow structures and textures are displayed and there is a striking agglomerate. At La Crête Point is the finest example of columnar jointing in the overlying Bouley Rhyolite Formation (shown above).

Gas bubble in volcanic breccia, seen between Anne Port bay and La Crête Point.

Flow fold in rhyolite at La Crête Point.

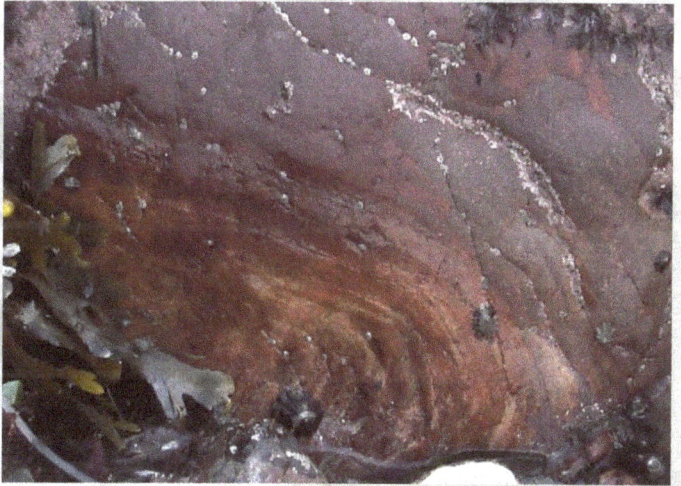

Local haematite mineralization occurs in a quartz vein in the rhyolite at La Crête Point.

22 - La Motte, Le Nez and Le Croc

SE Granite /Diorite/Loess	Palaeozoic/Pleistocene
SSI Status: International	Grid Ref: OS 6740 4606

Excellent outcrops of loess overlying diorite are seen at La Motte (Green Island). The picture above shows sections of loess (wind-blown silt) and glacial head (a freeze-thaw deposit). La Motte is part of a suite of loess sections related to human activity which includes Belcroute, Portelet and La Cotte de St Brélade. In addition, there are excellent outcrops of layered diorites, granite net-veining, as well as various dykes and fault styles at Le Nez and Le Croc.

Granite veins and brecciated-fractured diorite separated during intrusion illustrating net-veining features.

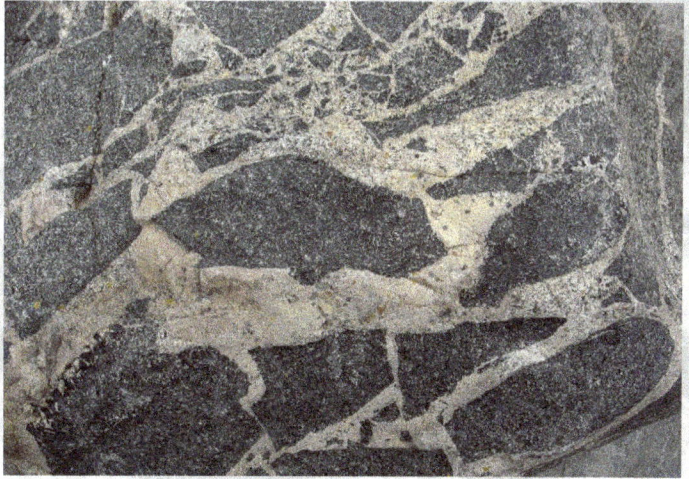

Granite veins have been intruded along joints into the diorite.
Inset picture: coarse hornblende and feldspar inclusion in fine diorite.

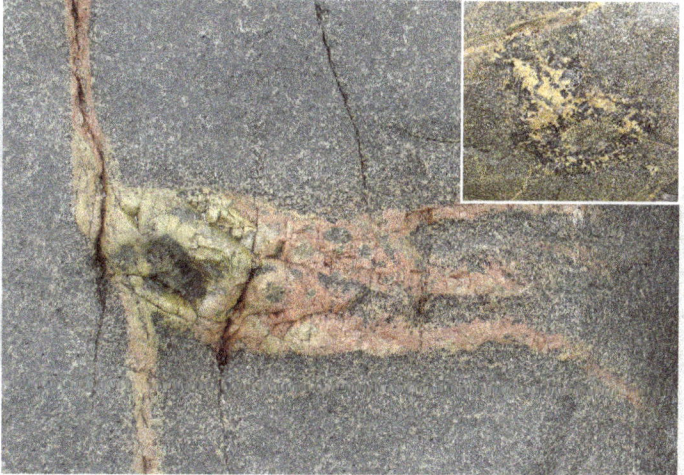

Dolerite dykes in the diorite form part of the Jersey Main Dyke Swarm.

SSI Boundaries

The following aerial images show the extent of the SSI boundaries for each of the geological sites described in this book. These sites are protected under provisions of the Planning and Building (Jersey) Law 2002 and it is an offence to disturb, damage or deface an SSI. This includes hammering, digging and the removal of rocks, sediment, plants and animals. Further information on the Planning and Building (Jersey) Law 2002 may be found at www.jerseylaw.je

1 - South Hill (scheduled for redesignation)

2 - Belcroute Bay

3 - Portelet Bay

4 - La Cotte de St Brélade

5 - St Ouen's Bay Peat (approximate)

6 - Le Mont Huelin Quarry

7 - Le Petit Étacquerel

8 - Le Grand Étacquerel

9 - Le Pulec

10 - Le Pinacle

11 - La Cotte à La Chèvre

12 - L'Île Agois, Crabbé

13 - Sorel Point

14 - Giffard Bay

15 - La Belle Hougue caves

16 - Les Rouaux

17 - Les Hurets, Bouley Bay

18 - L'Islet, Bouley Bay

19 - La Tête des Hougues

20 - La Solitude Farm

21 - Anne Port Bay

22 - La Motte, Le Nez and Le Croc

Glossary

Agglomerate. Volcanic rock consisting of angular volcanic fragments and sometimes of other rock types of different sizes and shapes set in fine-grained solidified volcanic ash.

Andesite. A finely crystalline, igneous volcanic rock of intermediate composition, a mixture of iron and magnesium minerals and feldspars.

Ankerite. A calcium, magnesium, manganese carbonate mineral. It is closely related to dolomite in its composition.

Biotite. A black to brown flaky mineral, in a subgroup of the Mica group.

Breccia. Rock made of coarse-grained, angular fragments of rock in a matrix of finer material.

Clast A rock fragment or grain resulting from the breakdown of any rocks.

Col. The lowest point or pass between two hills or peaks.

Columnar jointing. A series of generally hexagonal columns separated by vertical joints as a result of contraction during the cooling of a lava flow.

Concretion. A small, hard, compact mass of sedimentary rock formed by the concentration of minerals within the spaces between the sediment grains, usually spherical.

Conglomerate. A rock made of rounded pebbles or cobbles in a matrix of smaller grains.

Diorite. A coarsely crystalline, intrusive (plutonic) igneous rock of intermediate composition.

Dolerite. A dark-coloured igneous rock with medium size crystals, usually forming dykes and sills.

Dolomite. A carbonate mineral and rock composed of calcium magnesium carbonate.

Dyke. A narrow wall-like structure of igneous rock that formed at mid-crustal levels in joints or faults in a pre-existing rock body.

Eutaxitic. A banded appearance in hardened volcanic ash, resulting from the layering of glass shards and pumice (silica).

Fault. A fracture in the rock where adjacent sides have been displaced.

Feldspar. An alkali (potassium, sodium or calcium) silicate mineral, a most common igneous rock-forming mineral.

Felsite. A crystalline igneous rock made of quartz and feldspar; sometimes containing larger crystals and forming dykes and veins.

Fiamme. Pumice clasts inside an ignimbrite that have become stretched sideways while still hot as the deposit is compressed by the weight of overlying material.

Flow-banded rhyolite. A rhyolite flow that shows flow-banded textures formed as the lava flowed down the volcano.

Flow-banding. Rhyolite flow, often showing flow-fold structures that exhibit flow-banded textures formed as the lava flowed down the volcano.

Galena. The natural (often cubic) mineral form of lead sulphide. It is a widely distributed sulphide mineral and is the most important lead ore mineral.

Gabbro. A dark-coloured, plutonic intrusive igneous rock made of large crystals of iron and magnesium minerals.

Granite. A light-coloured plutonic intrusive igneous rock made of large crystals of feldspar and quartz minerals.

Granophyre. A variety of granite that contains intergrowths of quartz and feldspar.

Greywacke. A sandstone of quartz, feldspar and rock fragments (clasts) in a matrix of clay and silt.

Haematite. The most important ore mineral of iron. It is an oxide mineral that is found in a variety of colours (often red) and forms.

Head. Describes overlying glacial deposits (of sand, silt and angular rock fragments) at the very top of the cliffs, produced by soil flow after frozen ground has thawed.

Holocene. A geological time period extending from about 11,000 years ago to the present day.

Hornblende. A complex series of iron and magnesium minerals; not a mineral in its own right but is a general term to refer to the series.

Hornfels. A finely crystalline metamorphic rock with the hardness and texture of an animal horn, produced by thermal metamorphism of shale or mudstone.

Ignimbrite. A volcanic ash deposit consisting of fragments lava, and pumice which turns to glass when welded together by high temperatures to become banded in the lower part.

Intrusion. An igneous rock that has been formed under the Earth's surface from magma which has slowly melted the overlying rock or pushed into any cracks or spaces it can find.

Ipswichian Interglacial. The second to latest interglacial period of the last ice age between 130,000 to 110,000 years ago.

Iron pyrite. Also known as 'fool's gold', this is an iron sulphide made of iron and sulphur.

Joint. A fracture in the rock without any displacement of the adjacent sides, in contrast to a fault.

Lamprophyre. A dark-coloured dyke rock with large crystals surrounded by small crystals. Mica lamprophyre contains minerals from the Mica group.

Loess. A fine, silt-sized sediment which is formed by the deposition of windblown silt and clay from a glacial area.

Micas. A group of minerals with a sheet silicate structure, biotite and muscovite being the most common

Microgranite. A medium to microcrystalline intrusive plutonic igneous rock. It contains crystals which are smaller and more uniform than those of granite, indicating that the magma has cooled more rapidly.

Molybdenite. A mineral made of molybdenum and sulphur, a molybdenum sulphide.

Mudstone. A fine-grained sedimentary rock made of clay or mud.

Parataxitic. A exceptionally well-layered and drawn-out eutaxitic texture.

Periglacial. Refers to places around the edges of glaciated areas.

Playa. A clay pan, mud flat, or dried up lake.

Pleistocene. The geological Epoch which lasted from 1.6 million to 11,700 years ago, spanning the world's recent period of repeated glaciations and interglacials.

Pluton. A body of intrusive igneous rock that has crystallized from magma cooling deep below the surface of the Earth (a plutonic rock).

Porphyry. A variety of igneous rock consisting of large crystals dispersed in a finely crystalline matrix.

Quartz. An abundant mineral in many igneous, sedimentary and metamorphic rocks, and made of silicon and oxygen

Quaternary Period. A geological time Period between 1.6 million and the present day, consisting of the Pleistocene and Holocene Epochs,and is characterized by a series of glaciations and interglacials and the appearance of modern humans.

Rheomorphic. Used to describe flow structures that are only observed in high grade ignimbrites.

Rhyolite. A light coloured extrusive (volcanic) igneous rock, rich in silica, and made up of small crystals. The extrusive equivalent of granite.

Shale. A fine-grained, flakey (fissile) sedimentary rock made of mud that is a mix of flakes of clay minerals and tiny fragments of other minerals.

Siltstone. A fine-grained sedimentary rock made of silt-sized particles; coarser than shale and finer than sandstone.

Slump folds. These occur when layers of sediment move down a slope and fold or crumple.

Soil creep. The movement of soil down a slope under gravity.

Sphalerite. A mineral that is the main ore of zinc. It is usually found in association with galena, pyrite and other sulfides; sometimes referred to as zinc blende.

Spherulites. Small, usually spherical bodies consisting of radiating crystals and curved (concentric) bands, found in silica-rich volcanic rocks, e.g. rhyolites.

Stalagmite. A conical pillar in a limestone cave that is gradually built upward from the floor as a deposit of calcium carbonate from ground water seeping through and dripping from the cave's roof.

Strike-slip fault. A fault with a near-vertical plane where the adjacent walls move to the left or right past each other horizontally, with very little vertical motion.

Tombolo. A depositional landform between an island and the mainland, forming a narrow piece of land.

Turbidity current. A rapidly moving current, laden with sediment, flowing down a deltaic or continental slope.

Wolstonian. A middle Pleistocene glacial stage between 352,000 and 130,000 years ago that preceded the Ipswichian Interglacial.

Zinc blende. See *sphalerite*.

Picture Credits

The majority of the photographs were taken by Ralph Nichols during field studies for his website, www.jerseygeologytrail.net. Other photographs of La Belle Hougue Caves, La Cotte à la Chèvre, Le Pinacle, La Cotte de St Brélade and South Hill were taken by the following: Jeremy Percival for his website, prehistoricjersey.net; the UCL Research Archaeology Team re-studying La Cotte de St Brélade; Kevin McIlwee of Jersey Seasearch, John de Carteret, and Tim Liddiard of the Department of the Environment. Arthur Hill has also taken many other photographs, over which he has copyright, for the book *Jersey Geology Trail*.

www.ingramcontent.com/pod-product-compliance
Lightning Source LLC
Chambersburg PA
CBHW051438270326
41931CB00019B/3472